Working Paper Series
Congressional Budget Office
Washington, DC

The Outlook for U.S. Production of Shale Oil

Mark Lasky
Congressional Budget Office
mark.lasky@cbo.gov

May 2016
Working Paper 2016-01

To enhance the transparency of the work of the Congressional Budget Office (CBO) and to encourage external review of that work, CBO's working paper series includes both papers that provide technical descriptions of official CBO analyses and papers that represent original, independent research by CBO analysts. The information in this paper is preliminary and is being circulated to stimulate discussion and critical comment as developmental work for analysis for the Congress. Papers in this series are available at http://go.usa.gov/ULE.

The author wishes to thank Robert Arnold, Wendy Edelberg, Ronald Gecan, Joseph Kile, Jeffrey Kling, Kim Kowalewski, Jeffrey Werling, and seminar participants from CBO's Macroeconomic Analysis Division for helpful comments and suggestions. Jeanine Rees prepared the paper for publication.

Abstract

This paper models the production of energy from shale resources and examines the outlook for that production. Four key regions account for most U.S. oil from shale. Drilling activity in those regions is modeled as operating along an elastic supply curve, which has drifted down over time as mining productivity has improved. The elastic response of production to prices is gradual, making production insensitive to prices in the short run but quite sensitive within two to three years. That model, coupled with a forecast of crude oil prices that the Congressional Budget Office published in January 2016, indicates that the rig count, a simple measure of drilling activity, will probably fall until mid-2016. Thereafter, the gradual response of production to that decrease in drilling activity is expected to cause shale oil production to decline until mid-2017; the fastest rate of decline will probably occur in mid-2016. After 2017, however, continued improvements in mining productivity are expected to help production of shale oil regain its early-2015 peak by 2020 despite much lower prices than those in the 2013–2014 period. The large medium-term response of production to prices means that, absent a major supply shock, oil prices are unlikely to stray outside a range of $33 per barrel to $73 per barrel for a sustained period during the next five years.

1. Introduction

The Congressional Budget Office (CBO) forecasts the price of West Texas Intermediate (WTI), a type of crude oil, as part of its economic projections.[1] This paper uses that forecast to examine the outlook for U.S. production of crude oil from shale. CBO does not make a separate forecast of crude oil from shale, but understanding the outlook for that production can inform CBO's forecasts for WTI and for other variables, such as imports and business investment.

Crude oil from shale and similar formations, referred to collectively as tight oil, has had important effects on the U.S. economy over the past five years. Rising production of crude oil in four key regions (Bakken, Permian, Niobrara, and Eagle Ford), where almost all U.S. production of tight oil occurs, accounted for the dramatic increase in U.S. production of crude oil between 2010 and early 2015 (see Figure 1). That increase in production helped reduce the amount of crude oil that the U.S. imported from 9.3 million barrels per day during the first half of 2010 to 7.3 million barrels per day during the first half of 2015.

The outlook for U.S. production of tight oil is important because it affects global oil prices, the U.S. trade balance and GDP. Higher production directly boosts GDP and stimulates investment in capital used in the oil industry. The increase in U.S. production of crude oil probably contributed to the decline in global oil prices between 2011 and 2015. More recently, mining investment has fallen sharply and production of crude oil has slowed in response to a large drop in oil prices between mid-2014 and early 2016.

Two factors govern new production of tight oil: the price that drillers expect to receive when the oil is produced and mining technology. In this paper, the production of tight oil is modeled as a function of oil prices and technology so that production is projected based on a forecast of oil prices and technology. Although the amount of drilling activity is quite sensitive to the expected price of oil, both the response of drilling to the expected price and the response of total production to drilling are gradual. Consequently, a change in the price of oil induces a negligible response in total production over two to three months but a large response over two to three years. The productivity of mining has increased rapidly over the past five years and continues to grow.

A forecast for the one-year futures price of oil, based largely on the CBO's January 2016 forecast of the spot price of West Texas Intermediate (WTI), implies that the rig count in the four key regions will probably continue to decline until mid-2016 and then slowly recover. The response of total production to drilling is gradual, so total production of crude oil in those regions would continue to decline until mid-2017, eventually falling about 900,000 barrels per day from April levels. However, because mining productivity has grown rapidly and will probably continue to grow, production is expected to regain its early-2015 peak by late 2019, even though oil prices are forecast to remain well below the levels that brought production to its 2015 level.

---

[1] See Congressional Budget Office, *The Budget and Economic Outlook: 2016 to 2026* (January 2016), www.cbo.gov/publication/51129.

Figure 1.
U.S. Production of Crude Oil
Thousands of Barrels per Day

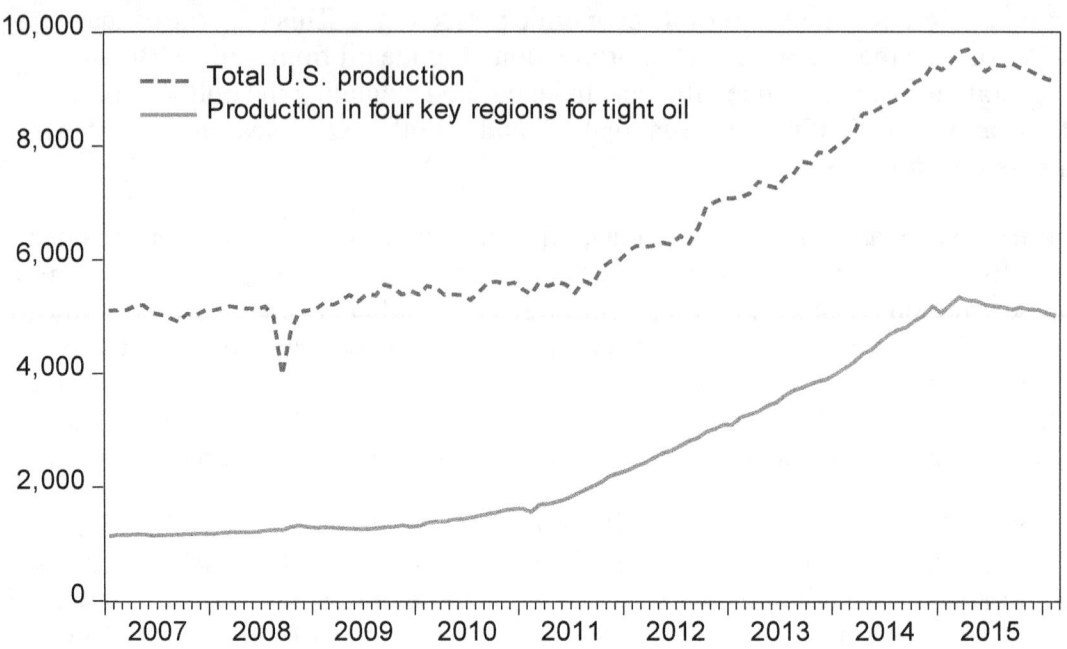

Source: U.S. Energy Information Administration.

Total U.S. production of crude oil can fall temporarily when a hurricane leads oil companies to temporarily halt off-shore production, as in September 2008.

The outlook for production beyond 2016 is quite sensitive to the forecasts for oil futures prices and mining technology. Production of crude oil in the four key regions at the end of 2018 would be 6 million barrels higher if the one-year futures price of oil were $80 per barrel from mid-2016 to 2018 than if the futures price were $30 per barrel over that period. Given the strong sensitivity of U.S. production to oil prices over two to three years, and if a major supply shock does not occur, spot prices for oil are unlikely to move outside a range of between $33 per barrel to $73 per barrel for a sustained period over the next five years.

Section 2 of this paper provides background on what tight oil is, where it is produced, and the steps in its production. Section 3 presents the data used in the analysis, which are drawn from the Energy Information Administration's (EIA's) *Drilling Productivity Report* (DPR).[2] Section 4 presents the model used in the paper as well as key theoretical concepts. Section 5 presents the empirical results for the rig count and for new production per rig. Section 6 discusses the history of the three components of new production per rig. Section 7 presents the base case forecast for oil production in the four key regions examined in this paper. Section 8 includes scenario analyses using alternative forecasts for oil prices and mining productivity.

---

[2] U.S. Energy Information Administration, *Drilling Productivity Report for Key Tight Oil and Shale Gas Regions* (April 2016), www.eia.gov/petroleum/drilling.

2. Background on the Production of Tight Oil

2.1. What is Tight Oil?

Tight oil is a light crude oil locked in certain rock formations, especially shale formations but also tight sandstone. It is often reached using hydraulic fracturing with horizontal drilling and other advances in drilling technology. That process (often called fracking) begins with drilling a vertical well to the depth of a shale formation and, from there, drilling a horizontal well into the formation. A high-pressure mixture of water, chemicals, and small particles is pumped into the well to create fractures in the formation, which are held open by the particles. Oil (and often gas) then flows into the well.[3]

One key difference between tight oil and conventional oil is that wells producing tight oil deplete much more rapidly than wells producing conventional oil. Globally, production from conventional oil fields declines by about 6 percent per year.[4] However, in the Eagle Ford region (a key region for production of tight oil in south Texas) from 2009 to 2013, monthly production of an average well declined by 64 percent to 70 percent during its first year in operation. That decline was even more dramatic after 2013 (see Figure 2). Decline rates differ by oil field but average at least 35 percent per year in each major producing region.

2.2. Where is Tight Oil Produced?

The seven key regions for the production of tight oil and shale gas (natural gas that comes from the same formations as tight oil) are Bakken, Niobrara, Permian, Eagle Ford, Haynesville, Marcellus, and Utica (see Figure 3). Most of the production from the Haynesville, Marcellus, and Utica regions is shale gas; this study focuses on the other four regions, which together produced nearly 5.2 million barrels of crude oil per day in 2015, or 97 percent of production in the seven regions (see Table 1).

2.3. Steps in the Production of Tight Oil

Timing of the production process is an important consideration when modeling how production of tight oil responds to oil prices. Producing oil and natural gas from shale involves several steps. The first is exploration. Once a promising site has been found and bought or leased, the driller must obtain a permit. Then, the driller must contract and schedule a drilling rig. Those steps can create lags between a change in the price of oil and a change in the number of drilling rigs in operation. Once a drilling rig is in place, it takes up to a month to drill a well. Once drilling is finished, it then takes roughly two months to case and complete a well and any associated infrastructure.[5] All of those steps are included

---

[3] See Congressional Budget Office, *The Economic and Budgetary Effects of Producing Oil and Natural Gas From Shale* (December 2014), pp. 3–4, www.cbo.gov/publication/49815.

[4] See International Energy Agency, *World Energy Outlook 2013* (November 2013), p. 463, www.worldenergyoutlook.org/weo2013.

[5] See U.S. Energy Information Administration, *Drilling Productivity Report: Report Background and Methodological Overview*, August 2014, p. 8, www.eia.gov/petroleum/drilling/pdf/dpr_methodology.pdf.

Figure 2.
Average Oil Production per Well in the Eagle Ford Region
Barrels per Day

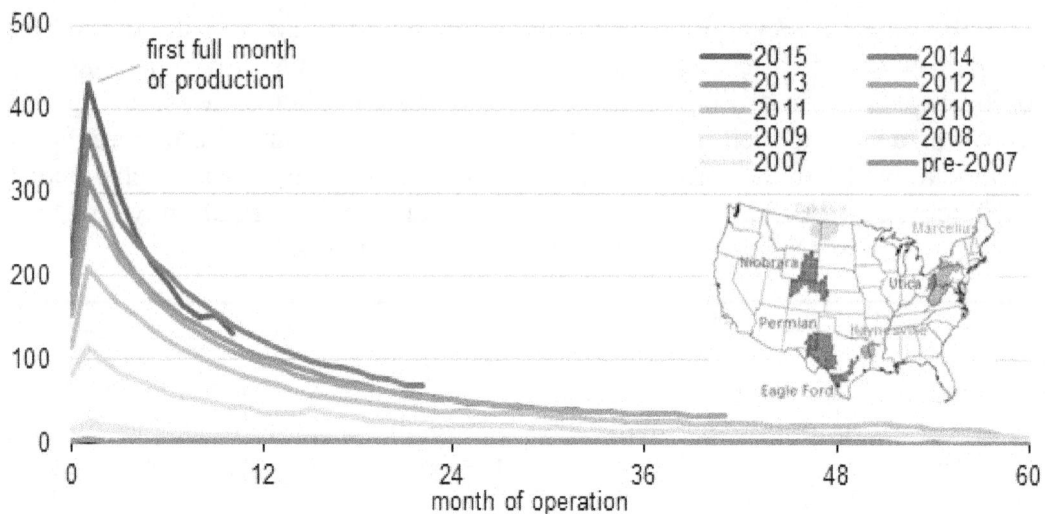

Source: U.S. Energy Information Administration, "Initial production rates in tight oil formations continue to rise," *Today in Energy* (February 11, 2016), www.eia.gov/todayinenergy/detail.cfm?id=24932.

Figure 3.
Key Tight Oil and Shale Gas Regions

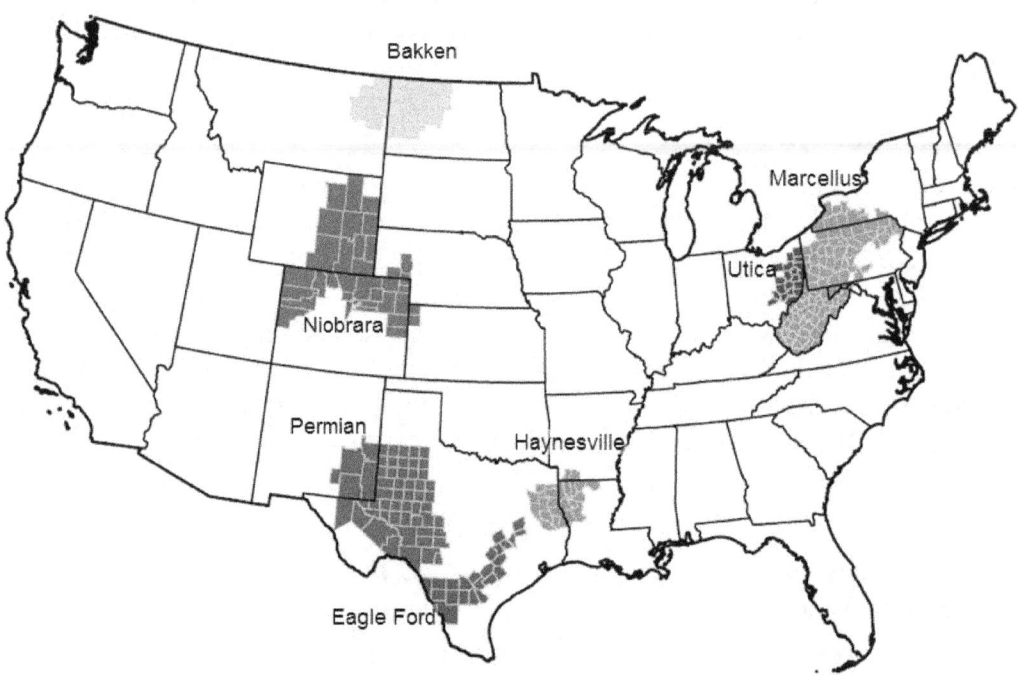

Source: U.S. Energy Information Administration.

Table 1.
Production per Day of Crude Oil in 2007, 2011, and 2015, by Region
Thousands of Barrels

Region	2007	2011	2015
Bakken	138	432	1,210
Eagle Ford	54	265	1,603
Niobrara	125	165	488
Permian	849	1,016	1,893

Source: U.S. Energy Information Administration.

in investment in mining structures—a component of GDP. Only after investment is complete does production begin. Because of the lags between changes in oil price and number of operating rigs, it takes new production three to nine months to respond to a change in the price of oil. The response of total production, which includes production from existing wells, to a change in the price is even more gradual.

The distinction between drilling and production is crucial when considering the response of oil production to oil prices. The Baker Hughes "rig count" frequently mentioned in the press is a measure of the number of active drilling rigs. Because a well does not even begin to produce oil or natural gas until two months after it has been drilled, there is no relationship between the rig count and production of oil in a given month. In fact, existing wells would continue to produce oil even if the rig count were to fall temporarily to zero. In short, rigs do not produce oil; rigs produce wells, and wells produce oil.

3. Data Used in the Analysis

The U.S. Energy Information Administration's *Drilling Productivity Report* is the primary source of data for this paper. The DPR contains monthly estimates, beginning in 2007, of the number of active drilling rigs; new production of oil and natural gas per rig active two months earlier; and overall production of oil and gas. The DPR focuses on the rig count rather than on the number of new wells because real-time data is more readily available for the rig count.

The data in the DPR is taken from several sources.[6] The monthly rig count is an average of weekly data from Baker Hughes. Initial production per rig is found by multiplying an estimate of wells per rig (based on the historical relationship between new wells and the rig count) by an estimate of new production per well in the first full month of production. According to data from Baker Hughes, wells drilled per month by an active rig in 2014 ranged from 1.25 in the Bakken region to 1.84 in the Eagle Ford region. Data for both the number of new wells and production per well in the first full month of production are from

---

[6] See U.S. Energy Information Administration, *Drilling Productivity Report: Report Background and Methodological Overview* (August 2014), www.eia.gov/petroleum/drilling/pdf/dpr_methodology.pdf.

DrillingInfo, Inc. The raw data for wells per rig and production per well vary greatly from month to month, so EIA smooths that data in order to obtain its estimates of new production of oil and gas per rig. Because production commences two months after drilling occurs, new production per rig is for rigs active two months earlier.

Data for production of oil and gas from existing wells also comes from DrillingInfo, Inc. Total production is the sum of production from new wells and from existing wells. Real-time data is not readily available for the number of wells, the production of new wells, or the production of existing wells. Consequently, estimates of total production in recent months are not based on actual data, but rather on past production, assumptions about the rate at which production from existing wells declines, the rig count, and expected production per well from new wells.

Although this paper focuses on oil, the DPR does not distinguish between oil and gas rigs. Because the productivity of active rigs is a key element in the analysis, I need a way of weighting the shares of oil and gas in new production. (The shares of oil and gas in new production differ from those in total production both because oil's share of new production has changed over time and because the depletion rates for oil and gas differ.) To do that, I combine production of oil and gas into a single fixed-weighted measure of energy using their average futures prices over 2009-2015 as weights. I use the price of WTI for oil and the price at Henry Hub, Louisiana for natural gas.[7] Equations fit better for a fixed-weighted index than for a chain-weighted index. By value, oil production has come to dominate total production in the four regions (see Figure 4).

4. Theory

The goal of the theory is to model production of tight oil as a function of technology and oil prices so that production can be projected based on assumptions about technology and CBO's forecast of oil prices. Modeling the production of tight oil involves several steps:

- Modeling the rig count as a function of oil prices and technology;

- Modeling new production per rig as a function of oil prices, technology, and the rig count;

- Estimating the depletion rate for new wells as a function of oil prices and technology; and

- Modeling total production as a function of new production and the depletion rate.

As discussed below, the term "technology" incorporates a few key theoretical concepts.

---

[7] The average year-ahead futures price of WTI during the 2009–2015 period was $86.60 per barrel in 2014 dollars, while the average year-ahead futures price of natural gas at Henry hub was $4.58 per million Btu in 2014 dollars, or $4.72 per thousand cubic feet. As an example of how oil and gas are combined, new production of oil in the four key regions was 216.9 thousand barrels per day in March 2016 and new production of natural gas was 499.1 thousand cubic feet per day. Total new production of energy per day in March 2016 was thus $86.60*216.9+$4.72*499.1, or $21,100 thousand per day in 2014 dollars.

Figure 4.
Crude Oil's Share of the Production of Crude Oil and Natural Gas in the Bakken, Eagle Ford, Niobrara, and Permian Regions
Percent of Value

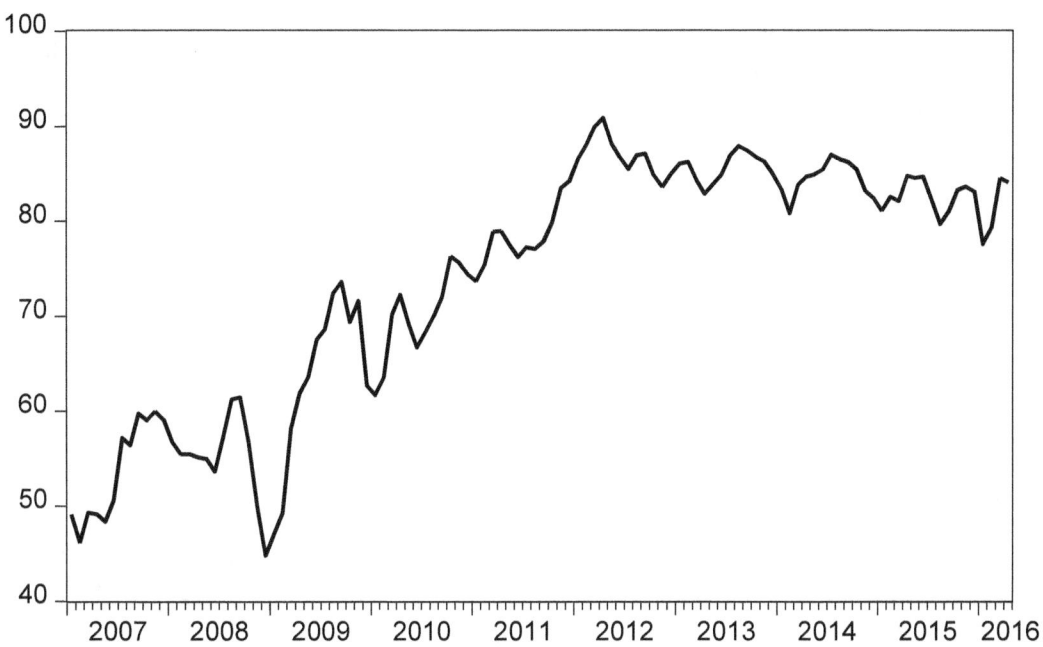

Source: Author's calculations, using data from the U.S. Energy Information Administration.

4.1. Key Theoretical Concepts

First-month production per rig (*fmp*) is modeled as the product of three concepts: energy produced per input, inputs per rig, and the depletion rate of new wells. Although a one-percent increase in any of those concepts boosts first-month production per rig by the same amount, those increases have different effects on total production over time.

The first of those concepts, energy produced per input (*epi*), measures mining productivity. For a rig operated for one month, *epi* is the total amount of oil and gas ultimately produced by the wells drilled by that rig divided by total inputs—the real cost of drilling, completing, and maintaining those wells.

The second concept, inputs per rig (*ipr*), measures the scale of production, such as the quality of rigs and the intensity of fracking of wells. For a given level of energy per input, an increase in inputs per rig boosts both energy produced and the cost of producing that energy by equal percentage amounts. Although energy produced per input governs the decision of whether or not to drill, inputs per rig govern how intensively an opportunity is exploited.

The third concept, the depletion rate of new wells (δ), measures the speed at which oil and gas is extracted from new wells. Assuming a geometric rate of depletion (i.e., a constant percentage rate of decline of production), first-month production per rig is proportional to

Figure 5.
Active Rigs and First-Month Production per Rig in Four Key Regions
Percentage Change, Year-Over-Year

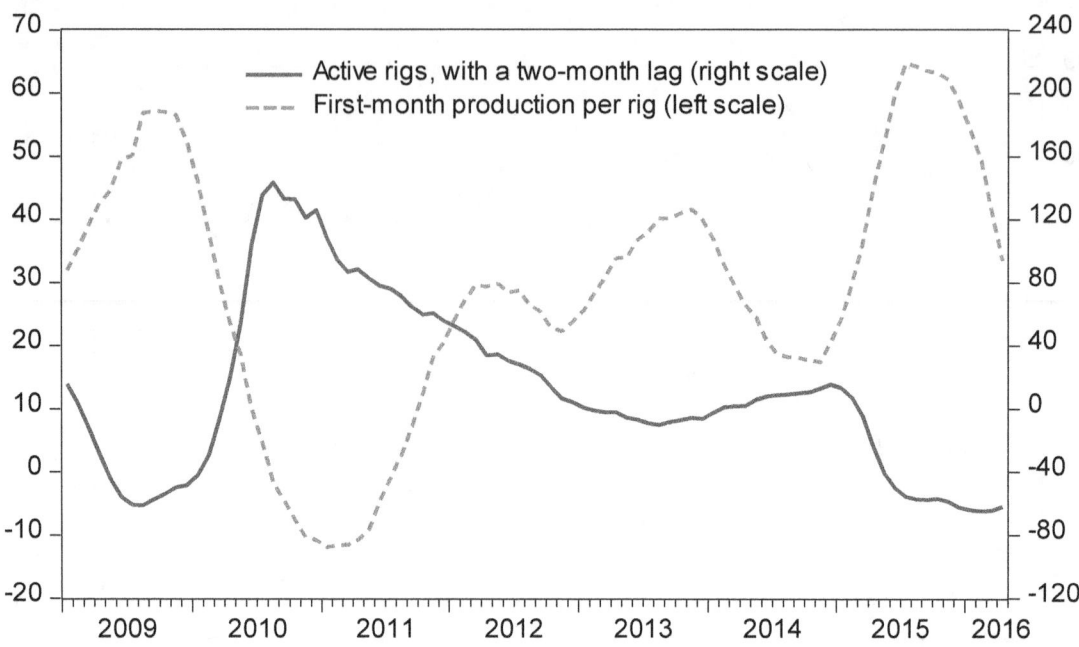

Sources: Author's calculations, using data from the U.S. Energy Information Administration.

the rate of depletion. A faster rate of depletion does not, by itself, boost the total amount of energy produced by a given well. Rather, it increases production in the first several months of a well's operation and reduces production in later months.

A single new technology may involve changes in all three determinants of first-month production per rig. For example, a new, more productive, technology (higher *epi*) may require more intense fracking (higher *ipr*) and lead to faster depletion rate (higher δ).[8] However, because the increases in *epi*, *ipr*, and δ have different effects on total production, it is necessary to model them separately.

Because not all investment opportunities are the same, two additional concepts are useful for the analysis: energy per input for a marginally profitable rig (*epim*); and energy per input for the 200th most profitable rig in service (*epi*200). Those concepts are useful because the data exhibit a strong inverse relationship between growth in the number of rigs in service and growth of first-month production per rig (see Figure 5). That inverse relationship is caused by drillers moving along a supply curve. For example, as oil prices

---

[8] For examples of improved technologies, see Tom DiChristopher, "Frackers Change Methods in 'Imploding' Oil Market," *CNBC* (September 2, 2015), www.cnbc.com/2015/09/02/frackers-change-methods-in-imploding-oil-market.html; and Trisha Curtis, *US Shale Oil Dynamics in a Low Price Environment*, Working Paper 62 (The Oxford Institute for Energy Studies, November 2015), pp. 5–6, 13.

fall, drillers forego their least profitable opportunities, pushing up both energy per input for a marginally profitable rig and average energy per input for rigs remaining in service. Although energy per input for the 200th most profitable rig in service is unobservable, it is a useful concept for measuring mining productivity. It removes the effects of changes in the rig count on average and marginal energy per input and reflects only the underlying technology. As the underlying technology of mining increases, $epi200$ increases proportionately.

4.2. The Rig Count and Energy per Input

Drillers move along the supply curve to the point where the expected value of the output from the wells drilled by an additional rig equals the cost of that rig and its associated inputs (such as labor and the costs of completing and maintaining the well). The nominal value of the output of the marginal rig is the price of its output, *peng*, times the energy produced by the marginal rig, which is energy per input for the marginal rig (*epim*) times inputs per rig (*ipr*). The nominal cost of a rig is inputs per rig times the nominal price index for those inputs, assumed to be the price index for GDP, *pgdp*. Although the deflator for mining investment might seem like a better choice for the price of inputs, the fit of the equations deteriorates significantly when that price is used instead of *pgdp*.

Ignoring the lags in the process of drilling and production, drillers employ rigs up to the point where expected marginal revenue equals expected marginal cost:

$$peng * ipr * epim = pgdp * ipr. \tag{1}$$

Cancelling out *ipr* from both sides and rearranging produces:

$$epim = \frac{pgdp}{peng}. \tag{2}$$

The greater the price of energy produced (*peng*), the lower the productivity (*epim*) required for a well to be profitable. That result accords with anecdotal evidence, such as a statement in November 2015 by the chief executive of DrillingInfo: "Right now, you are down to the best areas, the best rigs, the best people."[9]

Lags in the drilling and production process are important and must be considered when modeling drillers' decisions. Because of the two-month lag between when a well is drilled and when it begins to produce, the decision of how many rigs to employ in a particular month depends on the energy per input and inputs per rig expected for wells coming into service two months later, denoted $epim_{+2}$ and ipr_{+2}. (Drillers presumably know $epim_{+2}$ and ipr_{+2} when they drill.) Because of lags between the decision to drill and the commencement of drilling, the decision to drill depends on prices observed before drilling begins. Drillers base their decisions on futures prices because the incentive to drill depends on the price drillers expect to receive when oil is produced. Empirically, the best fit is obtained using a

---

[9] Bradley Olson and Erin Ailworth,"Low Prices Catch Up With U.S. Oil Patch," *Wall Street Journal* (November 21–22, 2015), pp. A1–A2.

seven-month moving average of the one-year-ahead real futures price, denoted *rpeng7*. That is, the number of rigs operating in the current month depends equally on real futures prices in that month and in each of the preceding six months.

After taking those lags into account,

$$epim_{+2} = \frac{1}{rpeng7}.$$ (3)

Energy per input for wells that will begin producing in two months depends inversely on a seven-month moving average of the real futures price of energy.

To link Equation 3 to the rig count, recall that drillers operate along a supply curve. Let *c1* be the elasticity of that supply curve, which is assumed to be constant over the entire curve. Because the rig count is positively related to the real price of energy and because energy per input for the marginal rig is inversely proportional to the real price of energy, the rig count (*rigs*) is inversely related to energy per input of the marginal rig:

$$\frac{rigs}{200} = \left(\frac{epim_{+2}}{epi200_{+2}}\right)^{-c1}.$$ (4)

The higher the rig count rises above 200, the less productive the marginal rig is compared to the 200th most profitable rig.

Substituting for *epim$_{+2}$* in Equation 4 using Equation 3 and rearranging the result produces an equation for the rig count as a function of energy prices and the technology of production:

$$rigs = 200 * (rpeng7 * epi200_{+2})^{c1}.$$ (5)

The rig count depends positively on the expected real price of energy and on underlying mining productivity.

4.3. Inputs per Rig

Inputs per rig (*ipr*) are assumed as a function of technological factors (*ipr_tech*) and the real price of energy:

$$ipr_{+2} = ipr\_tech_{+2} * rpeng7^{c2}.$$ (6)

A higher real price of energy can boost inputs per rig because diminishing returns to inputs per rig cause a higher price of energy to boost the marginal return to additional inputs. The technological factors affecting inputs per rig may differ from those affecting output per input, i.e., *ipr_tech* may differ from *epi200*.

4.4. The Rate of Depletion

The rate of depletion of new wells (*δ*) is also a function of technological factors (*δ_tech*) and the real price of energy:

$$\delta_{+2} = \delta\_tech_{+2} * rpeng7^{c3}. \tag{7}$$

The real price of energy can affect the depletion rate because when energy prices increase, longer-dated futures prices rise by less than shorter-dated futures prices. That difference creates an incentive to use wells that get oil out of the ground more quickly, which can be achieved by using technology that increases the rate at which oil can be extracted from a particular well.

Resources may also be redeployed to regions in which initial production is higher but wells are depleted more quickly. For example, wells drilled by one rig produce less in the first month of operation in the Permian region than in other regions but are depleted more slowly. Thus, lower near-term futures prices for oil reduce the incentive to drill in the Permian by less than in other regions. Lower oil prices probably contributed, at least in part, to making the Permian's share of rigs in the four regions higher in early 2016 than it had been since at least 2007 (when data began to be collected).

5. Empirical Results

5.1. Drilling Rigs

To estimate Equation 5, *epi*200 is modeled as a random walk with drift:

$$dlog(epi200_{+2}) = a_y + a_{y,11} + e_y \tag{8}$$

where *dlog* denotes the change in the logarithm, a_y is a constant, $a_{y,11}$ is a dummy variable equal to 1 through December 2011 to capture faster growth in that period, and e_y is an error term. To more precisely estimate the elasticity of supply, the logarithms use year-over-year changes instead of month-over-month changes. The estimated equation is:

$$dlog12(rigs) = c1 * 12 * a_y + c1 * sum12(a_{y,11})$$
$$+c1 * dlog12(rpeng7) + c1 * sum12(e_y) \tag{9}$$

where *dlog12* denotes the year-over-year change in the logarithm and *sum12* denotes a 12-month sum.

Rpeng7 is calculated as a seven-month moving average of the real price of one-year futures in WTI. Where those prices are unavailable, a linear combination of six-month and two-year futures prices is used. Adding the futures price for natural gas does not help empirically. Because this paper focuses on tight oil, the sample period begins in 2009 to avoid times when natural gas was a substantial portion of new production.

The demand for rigs is quite elastic with respect to the price of energy; estimated *c1* is greater than 2 (see Table 2). The estimates of a_y and $a_{y,11}$ indicate that productivity growth was rapid over the sample period, averaging 0.93 percent per month (12 percent per year) through 2011 and 0.43 percent per month (5.3 percent per year) thereafter.

Table 2.
Estimates for Drilling Rigs as a Function of Time Trends and the Futures Price of Oil

	Coefficient	T-Statistic
a_y	0.0043	5.0
$a_{y,11}$	0.0050	3.8
$c1$	2.15	32.7

Source: Author's calculations.

R-squared = 0.94.

The sample is monthly, from January 2009 to April 2016.

This table shows the results from estimation of Equation 9 for the rig count. Coefficients a_y and $a_{y,11}$ measure the effects of time trends. The coefficient $c1$ measures the effect of the futures price of oil.

Equation 9 fits the rig count well, as shown by the high r-squared and by the alignment of the actual rig count with the fitted data (see Figure 6).[10] The sharp decline in the rig count in 2015 can be explained by the sharp drop in the real futures price of oil in late 2014 and early 2015. Drilling fell because fewer prospective wells would make money at lower expected prices, not because of drilling companies' financial difficulties.

Given the close fit in 2015 and 2016 of an equation that does not include any financial variables, drillers' financial circumstances do not appear to play a significant role in determining rig count. Investors in the drilling sector are continuing to finance projects that will make money. However, even if bankruptcies do not affect the total rig count, bankruptcies will probably affect which drillers operate those rigs.

The behavior of the rig count as shown in Figure 6 is consistent with the gradual response of drilling to futures prices. For example, the real one-year-ahead futures price of WTI fell by 30 percent between September 2014 and December 2014, but the fitted rig count declined by just 15 percent over that period because of the gradual response of drilling to prices. In fact, the actual rig count declined by just 3 percent over those three months, indicating an even slower response of drilling to prices than assumed in the equation.

Using the results from Equation 9, annual supply curves for drilling activity can be constructed: each year's relationship between the rig count and the real futures price of oil (using a seven-month moving average). To construct annual supply curves, Equation 5 is rearranged to obtain year-by-year estimates of $epi200_{+2}$, and then Equation 5 is used to solve for $rigs$ as a function of $rpeng7$ using that $epi200_{+2}$. The results are shown in Figure 7.

---

[10] Because Equation 9 is estimated in differences and not levels, the scaling of the fitted rig count in levels is arbitrary. The fitted rig count is scaled such that its average equals the average of the actual rig count over the sample period. The fitted line is thus $148.78 * exp[c1 * (a_y * t + a_{y,11} * t11)] * rpeng7^{c1}$, where t is 1 in January 2005 and increases by 1 per month, $t11$ equals t through December 2011 and is constant thereafter, and $rpeng7$ is scaled to average to 1.0 in 2014.

Figure 6.
Rig Count, Actual and Fitted, in Four Key Regions
Average Number of Active Rigs per Month

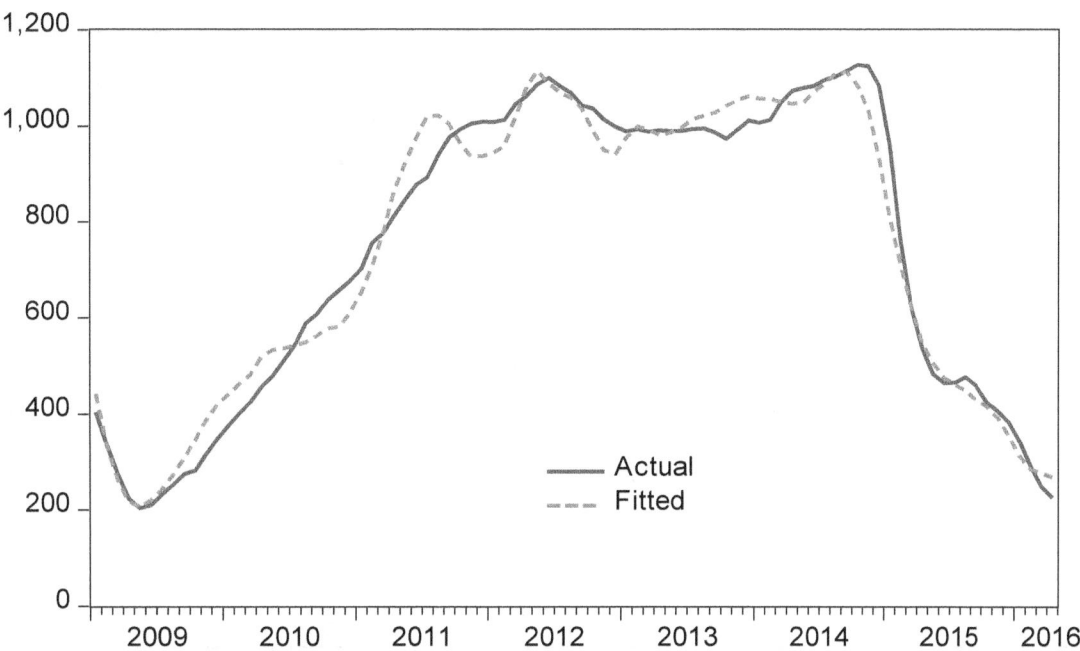

Sources: Author's calculations, using data from the U.S. Energy Information Administration.

The supply curve has shifted down over time as mining productivity ($epi200_{+2}$) has improved every year. The larger differences between the curves through 2012 reflect the faster productivity growth estimated during that period. (Although productivity growth slowed early in 2012, the higher growth in 2011 affects the year-over-year comparison in 2012.) The supply curve has shifted down far enough over time that the rig count in the four regions was slightly larger in 2015 than in 2010 even though the seven-month moving average of the real futures price of oil was almost 31 percent lower. Drillers used improvements in mining productivity to offset the adverse effect on profitability of lower oil prices.

Figure 7 provides visual evidence that the supply curve for drilling activity is flat, meaning that the number of rigs in operation is very sensitive to the real futures price of oil. If the supply curve were less elastic, making the supply curve steeper, the reductions in the rig count in 2013 and especially 2015 would be difficult to explain. A less elastic supply curve would have to shift upward in 2013 and 2015 to be consistent with the rig count.

5.2. First-Month Production per Rig

First-month production per rig, which occurs two months after drilling, is the product of energy per input, inputs per rig, and the depletion rate:

$$fmp_{+2} = epi_{+2} * ipr_{+2} * \delta_{+2}. \tag{10}$$

Figure 7.
Supply Curves for Drilling Activity in Four Key Regions
2014 Dollars per Barrel

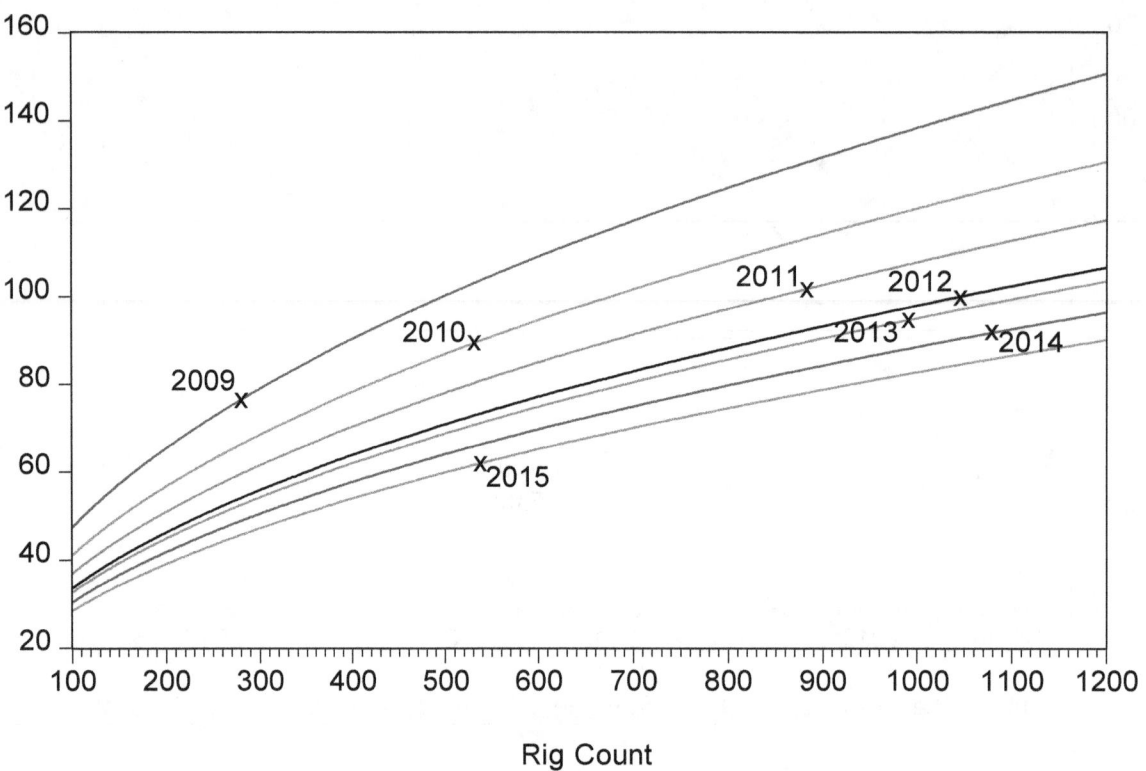

Rig Count

Sources: Author's calculations, using data from the Energy Information Administration, CME Group, the *Wall Street Journal*, and the Bureau of Economic Analysis.

The futures price of oil is the annual average of 7-month moving averages of the 12-month futures price per barrel of West Texas Intermediate crude oil in 2014 dollars per barrel. The rig count is the annual average of active rigs per month.

The equation for *epi* (derived in the appendix) shows that:

$$epi_{+2} = \frac{1}{1-1/c1}\, epi200_{+2} \left(\frac{rigs}{200}\right)^{-1/c1}. \tag{11}$$

As the rig count rises and less productive wells are added, average energy per mining input epi_{+2} declines for a given level of technology $epi200_{+2}$. Substituting for $epi200_{+2}$ using Equation 8 and taking logarithms and twelfth differences, the result is:

$$dlog12(epi_{+2}) = 12 * a_y + sum12(a_{y,11}) - \frac{1}{c1} dlog12(rigs) + sum12(e_y). \tag{12}$$

To model *ipr* and *δ*, *ipr_tech* is expressed as:

$$dlog(ipr\_tech_{+2}) = a_r + a_{r,11} + e_r \tag{13}$$

14

Table 3.
Estimates for First-Month Production per Rig as a Function of Time Trends, the Rig Count, and the Futures Price of Oil

	Coefficient	T-Statistic
$a_y + a_r + a_\delta$	0.0233	19.8
$a_{y,11} + a_{r,11} + a_{\delta,11}$	-0.0013	-0.9
$c1$	2.15	8.1
$c2 + c3$	0.464	3.7

Source: Author's calculations.

R-squared = 0.82.

The sample is monthly, from January 2009 to April 2016

This table shows the results from estimation of Equation 15 for first-month production per rig. Coefficients $a_y + a_r + a_\delta$ and $a_{y,11} + a_{r,11} + a_{\delta,11}$ measure the effects of time trends. The coefficient $c1$ measures the effect of the rig count and $c2 + c3$ measures the effect of the futures price of oil.

and δ_tech is expressed as:

$$dlog(\delta\_tech_{+2}) = a_\delta + a_{\delta,11} + e_\delta \qquad (14)$$

where variables are analogous to those in Equation 8. Substituting those expressions into Equations 6 and 7 and combining with Equation 12, the year-over-year change in first-month production is:

$$dlog12(fmp_{+2}) = 12 * (a_y + a_r + a_\delta) + sum12(a_{y,11} + a_{r,11} + a_{\delta,11})$$
$$- \frac{1}{c1} dlog12(rigs) + (c2 + c3)\, dlog12(rpeng7) + sum12(e_y + e_r + e_\delta). \qquad (15)$$

Growth of first-month production depends on the technological factors governing *epi*, *ipr*, and δ, on the real futures price of energy, and on the rig count.

Results of estimating Equation 15 are shown in Table 3. Estimated coefficients on the rig count and energy prices have the correct sign and are highly significant. One noteworthy result is that the estimate of *c1* is very close to that obtained above when estimating a different equation not containing first-month production. The negative coefficient on $a_{y,11} + a_{r,11} + a_{\delta,11}$ implies that growth of first-month production per rig accelerated slightly after 2011 despite the deceleration of the growth of mining productivity estimated above.

With Equation 15, the wild fluctuations in the growth of first-month production per rig, seen in Figure 5, make sense. The largest fluctuations are attributable to movements along

15

Figure 8.
First-Month Production per Rig
Percentage Change, Year-Over-Year

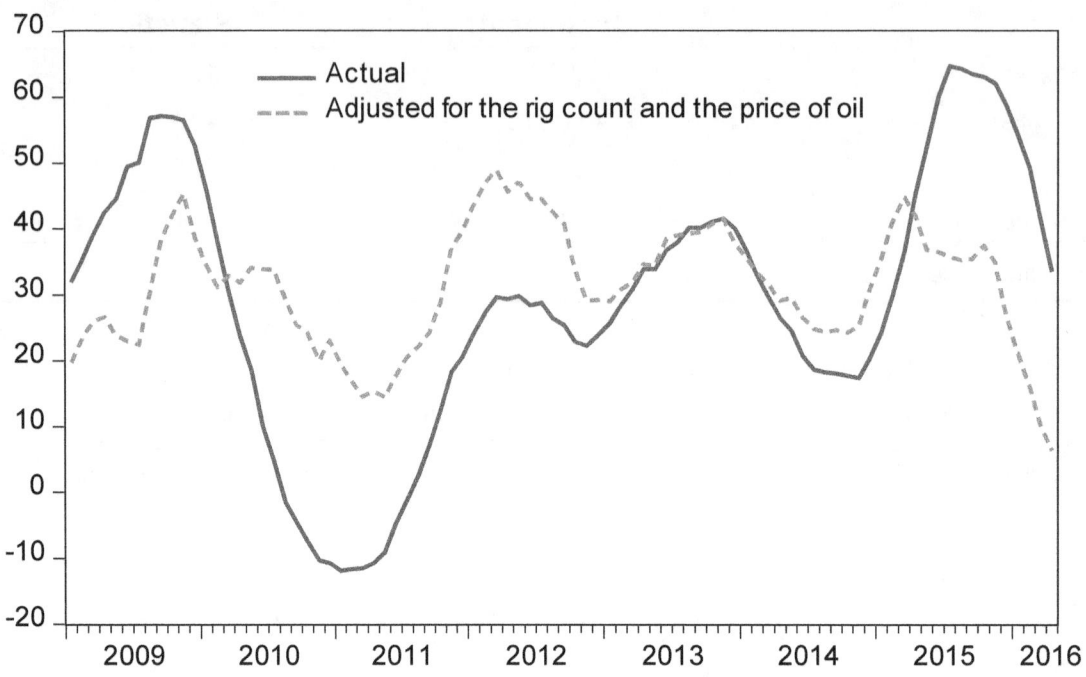

Source: Author's calculations, using data from the U.S. Energy Information Administration.

the supply curve and to changes in inputs per rig and depletion rates stemming from changes in energy prices (see Figure 8).[11] For example, the fall in first-month production per rig during 2010 and early 2011 reflected a movement up the supply curve—as the rig count soared and comparatively less productive wells were drilled—rather than a decline in the underlying productivity of mining. Similarly, the surge in first-month production per rig in 2015 reflected a move back down the supply curve.

6. History of the Components of First-Month Production per Rig: Energy per Input, the Depletion Rate, and Inputs per Rig

To forecast the production of tight oil, a forecast of the components of first-month production per rig (*fmp*) is needed: energy per input (*epi*), inputs per rig (*ipr*), and the depletion rate (*δ*). That forecast requires the history of those components. An estimate of *epi* can be backed out of the equation for the rig count. Dividing *fmp* by estimated *epi* produces an estimate of *ipr* times *δ*. Unfortunately neither of those can be observed directly. The depletion rate for all production is observed instead of the depletion rate for

---

[11] In Figure 8, first-month production per rig adjusted for the rig count and the price of oil is $fmp *$
$\left(\frac{rigs_{-2}}{200}\right)^{1/c1} rpeng7_{-2}^{-c2-c3}$ using parameter estimates from Table 3.

new production. The time series for δ is a simple function of *ipr* times δ and broadly reproduces the depletion rate for all production.

6.1. Energy per Input

A mathematical expression for energy per input for the 200th most-profitable rig (*epi*200, a measure of mining productivity independent of the rig count) can be found by inverting Equation 5 to obtain:

$$epi200 = \left(\frac{rigs_{-2}}{200}\right)^{1/c1} * \frac{1}{rpeng7_{-2}}. \tag{16}$$

For a given oil price, a greater rig count indicates higher productivity because greater productivity makes more rigs profitable. For a given rig count, a lower price of oil indicates higher productivity because a lower price of oil must be offset by higher productivity to keep the rig count from falling.

The expression for *epi*200 assumes that the rig count is affected only by mining productivity and the real price of oil and that the lag structure is constant over time. Other factors, such as financing constraints and the weather, could also affect the rig count, and the lag structure may change over time. In Equation 8, which specifies *epi*200 in terms of time trends and an error term e_y, those other factors are captured in e_y.

Thus, *epi*200 can be expressed using either Equation 16 or the time trends in Equation 8. Figure 9 shows the time paths for those two specifications of *epi*200. The unusually slow response of the rig count to lower oil prices in late 2014 and early 2015 appears as a temporary surge in productivity in the path estimated using Equation 16. Given that such a temporary surge of actual productivity is implausible, *epi*200 is estimated using the smooth line in Figure 9, calculated using Equation 8. *Epi* is then calculated by inserting that estimate into Equation 11. Dividing *fmp* by that estimate of *epi* yields an estimate of *ipr*$\ast\delta$.

6.2. Depletion Rates

Although the product of inputs per rig (*ipr*) and the depletion rate of new production (δ) can be estimated using the method shown in the previous section, neither of those variables can be estimated separately. The only data available is the depletion rate for all production, which itself is noisy. A strategy to estimate those variables separately is to find a time series for δ that is a simple function of *ipr*$\ast\delta$ and that broadly reproduces the depletion rate for all production.

Denoting total production as *prod*, new production as *new*, and using the subscript -1 to denote a one-month lag, the monthly depletion rate for existing production is:

$$-\frac{prod}{prod_{-1}} + 1 + \frac{new}{prod_{-1}}. \tag{17}$$

The monthly data is erratic but shows a clear upward trend over time (see Figure 10). Large monthly swings, such as those in late 2008 and early 2011, probably reflect a temporary shutting in of production followed by a resumption of production. Depletion in the most

17

Figure 9.
Energy per Input, Adjusted for Rig Count
Index, 2014 = 1.0

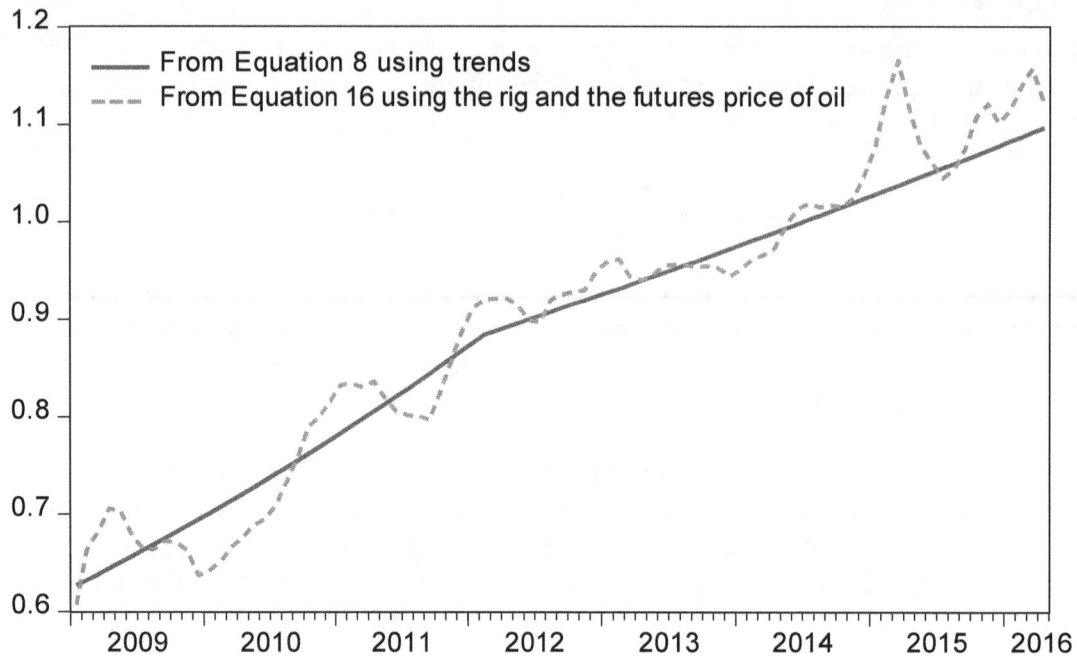

Source: Author's calculations, using data from the U.S. Energy Information Administration.

recent months is unusually smooth because those are not actual data but rather EIA's projections.

To determine a path for the depletion rate of new production, the effect of that depletion rate on the depletion rate for all production needs to be determined. The depletion rate for all production depends on the depletion rates of new production in preceding months. To be exact, that would require keeping track of the depletion rate for new production in each prior month. The appendix shows the simpler procedure used to keep track of those depletion rates.

Adjusting for the impact of futures prices for oil, a reasonable fit for the total depletion rate is obtained by assuming that increases in δ account for 80 percent of the growth of $ipr*\delta$ during the 2008–2011 period and for 25 percent of growth thereafter. The effect of oil prices on ipr ($c2$) is also assumed to be half as large as the impact of oil prices on δ ($c3$). Further details of the calculation are discussed in the appendix.

The resulting estimated depletion rate for all production of oil is shown in Figure 11. The depletion rate increased most rapidly in late 2011 and early 2012 as a large amount of new production with high depletion rates came online. After rising steadily since 2008, the estimated depletion rate declined slightly in the second half of 2015 and early 2016 as production shifted toward the Permian region, with its relatively low depletion rates. In the

Figure 10.
The Combined Monthly Depletion Rate for Oil and Natural Gas in Four Key Regions
Percent

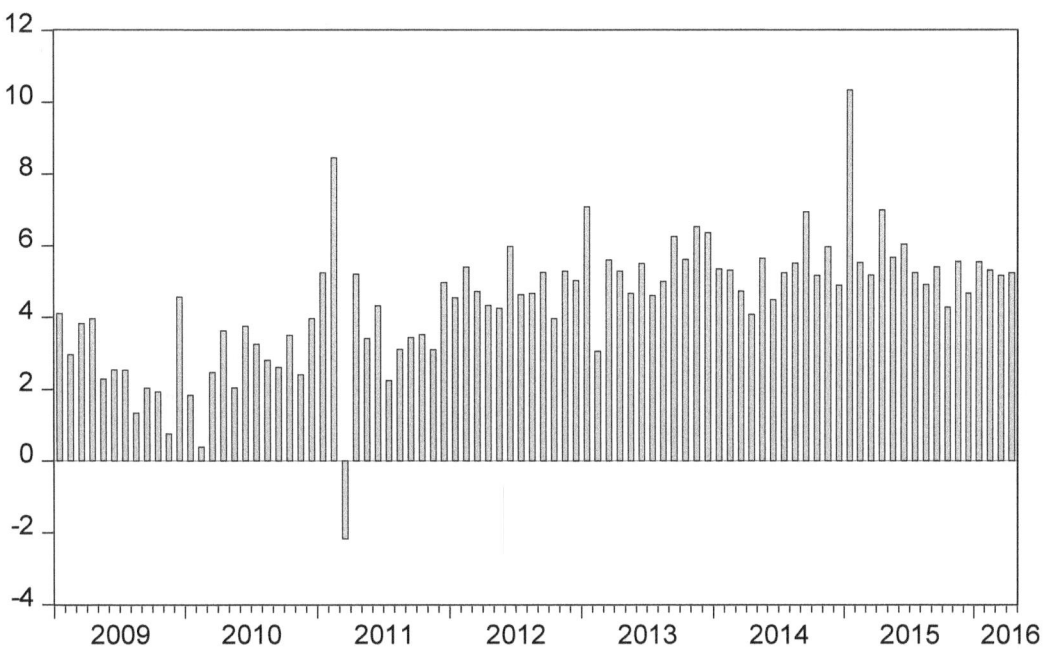

Source: Author's calculations, using data from the U.S. Energy Information Administration.

The four key regions are Bakken, Eagle Ford, Niobrara, and Permian.

model, that shift is captured through the coefficient $c3$. The depletion rate for natural gas has risen much more slowly than the depletion rate for oil.

6.3. Inputs per Rig and Energy per Rig

The estimate of inputs per rig (*ipr*) equals the estimate of *ipr*$*\delta$ divided by the estimate of the depletion rate for new production (δ). Because growth of δ accounts for 80 percent of the growth of *ipr*$*\delta$ through 2011 but for just 25 percent thereafter, growth of *ipr* was modest through 2011 and strong thereafter. That finding is consistent with the upward trend in the ratio of real private investment in mining exploration, shafts, and wells for petroleum and natural gas to the overall U.S. rig count beginning in late 2011. Strong growth of inputs per rig after 2011 is also consistent with a sharp increase in pounds of fracking proppant per well since mid-2012.[12] (A proppant is a solid material, typically sand, used to keep a hydraulic fracture open.)

Figure 12 apportions the growth of first-month production per rig (*fmp*), adjusted for the rig count and the price of oil, into its three components. Growth of adjusted *fmp* (the solid line in the figure) is identical to the red dashed line in Figure 8. Removing the growth rate of

---

[12] Dan Murtaugh and Devid Wethe, "Why America's Oil Output Refuses to Collapse," *Bloomberg Business* (September 30, 2015), http://tinyurl.com/pupk5o8.

Figure 11.
The Combined Monthly Depletion Rate for All Production of Oil in Four Key Regions
Percent

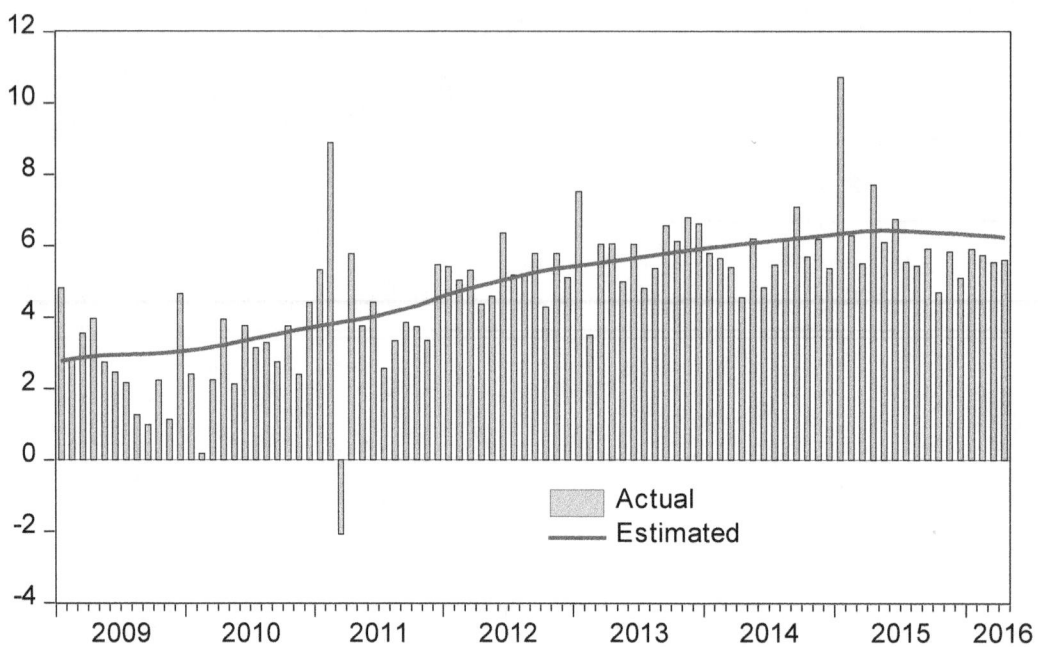

Source: Author's calculations, using data from the U.S. Energy Information Administration.

The four key regions are Bakken, Eagle Ford, Niobrara, and Permian.

depletion produces the dashed line—the growth of the total amount of energy that a rig will ultimately produce, or energy per rig. A higher depletion rate, by pulling production forward, boosts first-month production per rig without boosting the total amount of energy produced per rig. The lower dashed line in Figure 12 shows the contribution to the growth of energy per rig from productivity growth in the drilling sector, or the growth of energy per input. The difference between the two dashed lines reflects growth of inputs per rig. That captures such things as more expensive rigs and more intense fracking, which generate increased production per rig because of increased inputs.

7. Base Case Forecast

A forecast of the production of tight oil requires forecasts of the real futures price of oil and of the components of first-month production per rig. Forecasts of the rig count and production of oil are functions of those variables. The last historical data from the DPR is for April 2016, so the forecast begins in May 2016.

7.1. Forecast of the Futures Price of Oil

Although the CBO does not forecast futures prices of WTI, it does forecast the spot price of WTI. CBO projects that spot price to rise from $40 per barrel in the second quarter of 2016 to $46 per barrel in the second quarter of 2017 and then by between $3 per barrel and $5 per

Figure 12.
First-Month Production per Rig and Its Components, Adjusted for Rig Count and Price of Oil
Percentage Change, Year Over Year

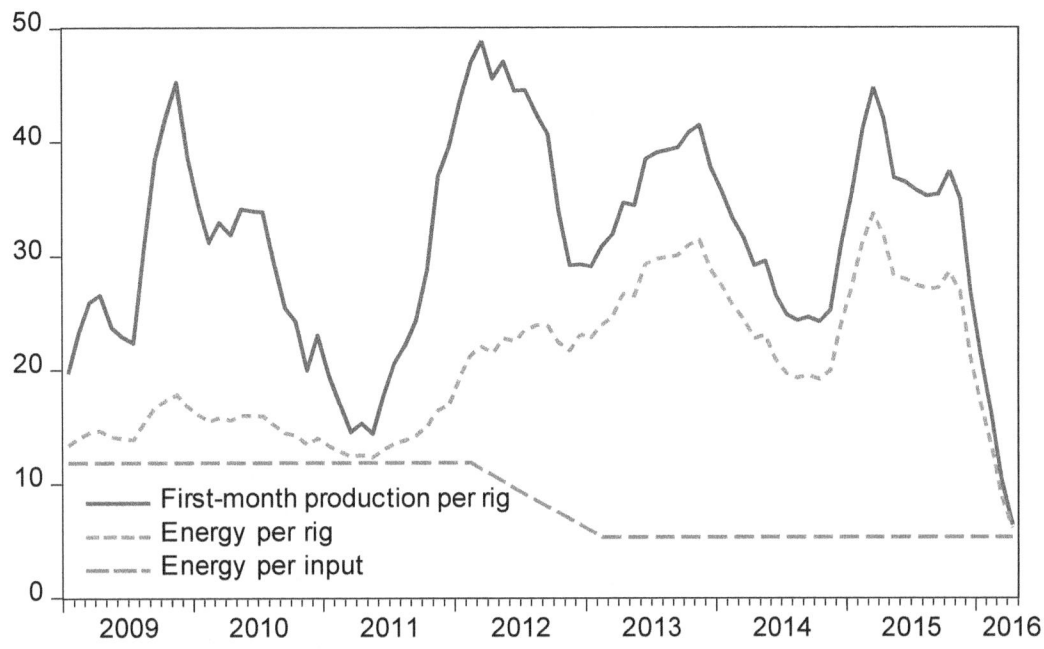

Source: Author's calculations.

barrel per year through 2021. The one-year futures price is assumed to equal the year-ahead forecast of the spot price. The result is a path for the one-year futures price that starts at $46 per barrel in May 2016 and increases by $3 to $5 per year to $62 per barrel (or $56 per barrel in 2014 dollars) at the end of 2020. A model of oil prices is beyond the scope of this paper, so whether CBO's forecast for spot prices is consistent with U.S. production of tight oil in the base case cannot be determined. Data for *pgdp*, the price index used to convert the price of oil into 2014 dollars, are from CBO's January 2016 forecast.

7.2. Forecast of the Components of First-Month Production per Rig

The rate of improvement of energy per input will probably slow in the future for two reasons. First, as the technology of hydraulic fracturing matures, further gains will probably become more difficult to achieve. Second, drillers move along a supply curve, exploiting the fields they think are the best available and only drilling in less productive fields as the price of oil rises. As the most productive fields are gradually pumped dry, more resources are needed for a given technology to produce the same amount of oil. Consequently, growth of energy per input of the 200th most profitable rig (*epi*200) is assumed to continue at its post-2011 annual rate of 5.3 percent through mid-2016 and then gradually decline by 0.075 percentage points per month (0.9 percentage points per year) to 1.3 percent by the end of 2020 (see Figure 13).

Figure 13.
Base-Case Forecast of Energy Produced per Rig, Adjusted for Rig Count and Price of Oil
Percentage Change, Year Over Year

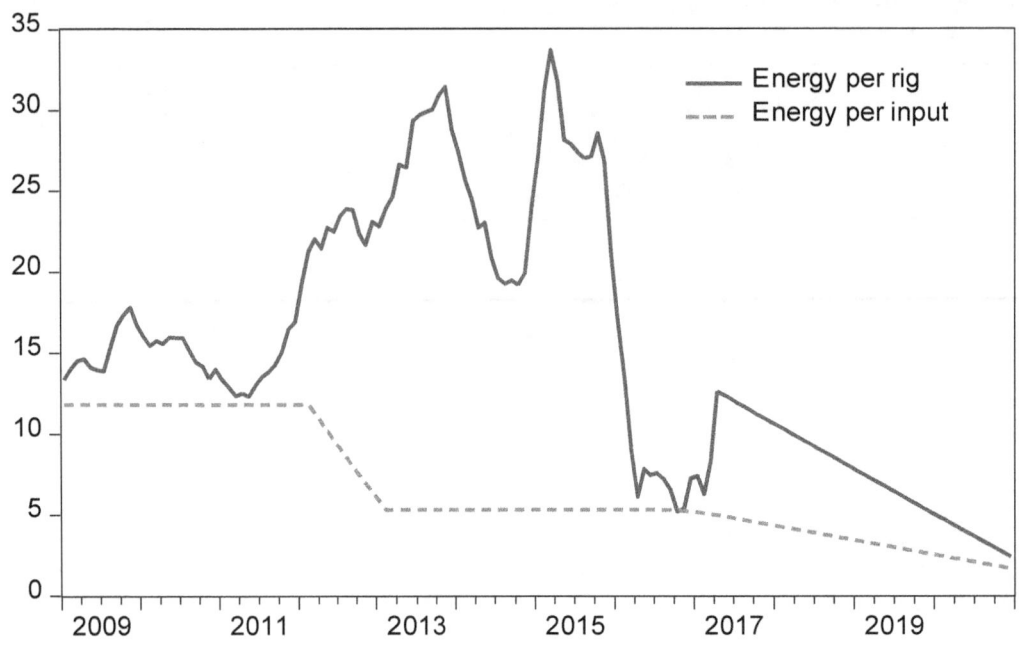

Source: Author's calculations.

Because adjusted inputs per rig (*ipr_tech*, the distance between the two lines in Figure 13) have varied so much over history, their forecast is one of the most uncertain elements of the base case forecast. The growth of *ipr_tech* is expected to slow for the same reasons as *epi*200. The growth rate of *ipr_tech* starts at 8 percent, somewhat below its historical average, and is then reduced by 0.15 percentage point per month—twice as rapidly as for *epi*200. *Ipr_tech* has less of an effect on oil production than *epi*200 because *ipr_tech* and *epi*200 both affect first-month production per rig but *ipr_tech* does not affect the rig count.

To forecast the depletion rate for new production (δ), δ_tech is assumed to continue to account for 25 percent of the growth of $\delta\_tech*ipr\_tech$, as it has since 2012. The growth rate of δ_tech is thus one-third the growth rate of *ipr_tech*. Because the depletion rate alters the timing rather than the amount of production, it has little effect on the base case.

7.3. Base Case Forecasts of the Rig Count and New Production of Oil

Given forecasts for the real futures price of oil and *epi*200, the rig count (*rigs*) can be forecast using Equation 5, $rigs = 200 * (rpeng7 * epi200_{+2})^{c1}$.

In light of recent data for *rigs* below fitted values, *rigs* is reduced by an amount that starts at 50 in May 2016 and June 2016 and shrinks by 15 percent per month thereafter. Given forecasts for the rig count, the real futures price of oil, and the exogenous drivers of technology (*epi*200, *ipr_tech*, and δ_tech), Equations 6, 7, and 11 determine *ipr*, δ, and *epi*—the components of first-month production (*fmp*). Multiplying first-month production

22

Figure 14.
Base-Case Forecasts of the Rig Count and New Production of Tight Oil in Four Key Regions
Monthly Rates

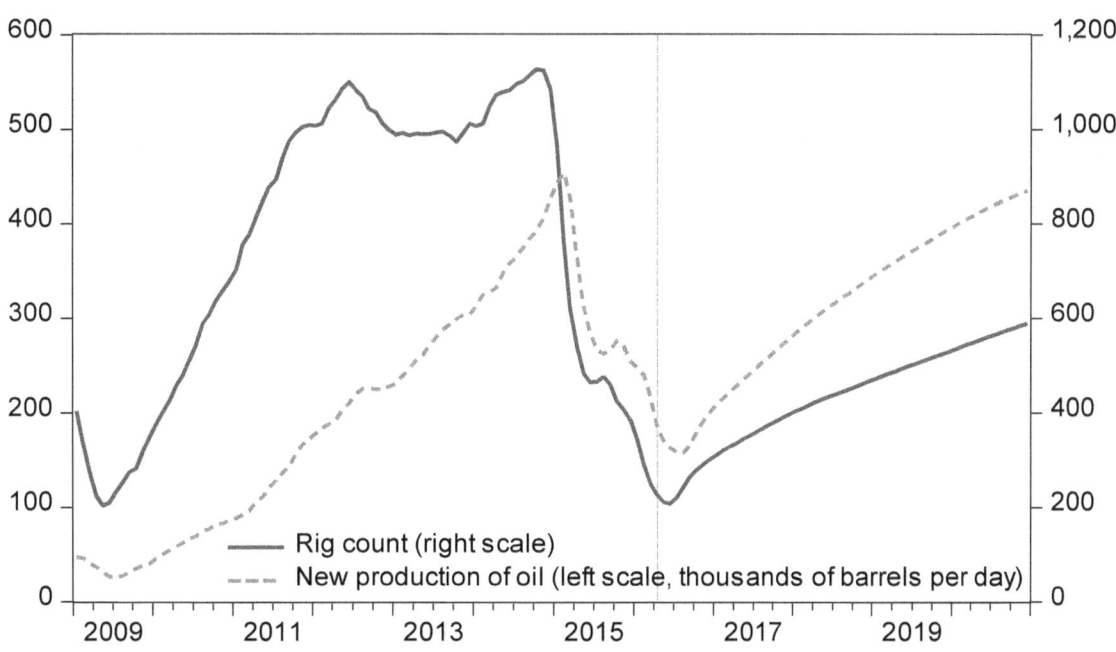

Source: Author's calculations.

The four key regions are Bakken, Eagle Ford, Niobrara, and Permian.

by the rig count lagged two months yields new production of energy. Oil's share of new production by value is assumed to remain near its recent level of about 90 percent.

The resulting forecast shows the rig count in the four main regions for oil production declining from 226 in April 2016 to about 210 per month in May and June before beginning a gradual recovery (see Figure 14). The rig count has continued to decline since the low point for futures prices in January because of the lagged effects of the price on the rig count. Because of the time between the decision to drill and the commencement of drilling, futures prices in January continue to affect the rig count for several months.

The base case shows the rig count recovering slowly beginning in July 2016, as oil prices slowly rise and energy per input continues to improve. Nevertheless, the rig count is expected to remain well below its 2014 levels—assuming that oil futures prices remain well below their 2014 level.

The base case for new production follows a similar pattern to that of the rig count but with faster growth. That faster growth is attributable to growth of first-month production per rig, which increases by 17 percent between April 2016 and December 2020 in the base case, or by 49 percent adjusted for changes in the rig count and oil prices. Despite a rig count that remains subdued by 2011–2014 standards, new production of tight oil is only 4 percent below its February 2015 peak by the end of 2020 in the base case.

Figure 15.
Base-Case Forecast for Production and Depletion of Tight Oil in Four Key Regions
Thousands of Barrels per Day

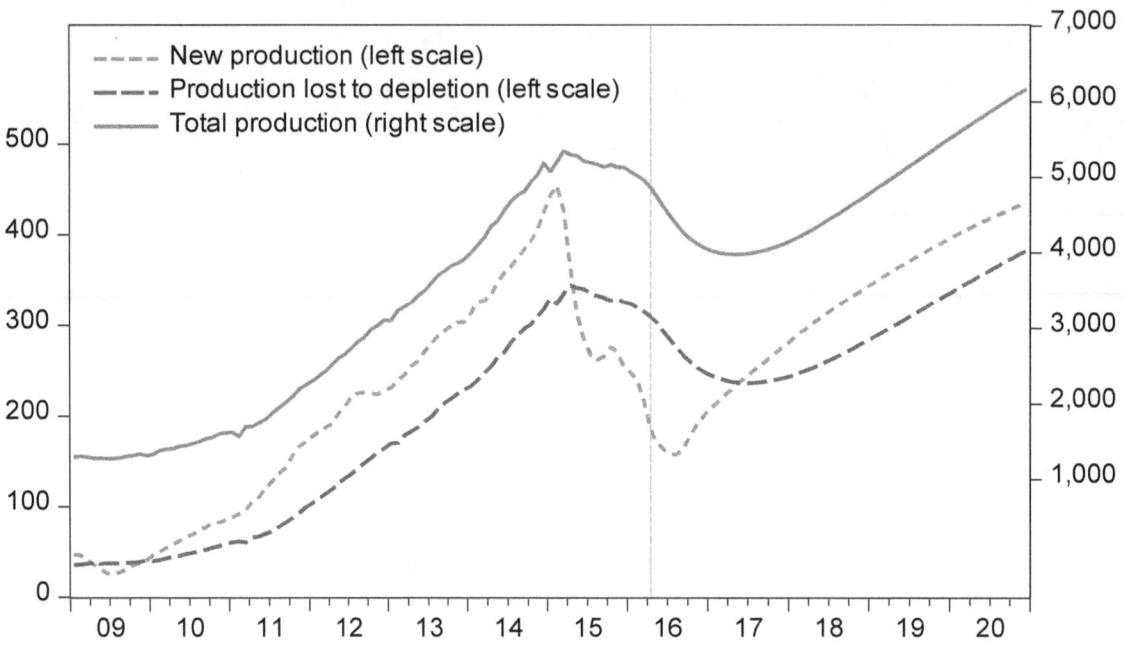

Source: Author's calculations.

The four key regions are Bakken, Eagle Ford, Niobrara, and Permian.

7.4. Base Case Forecast of the Production of Tight Oil

Total production of tight oil in a given month equals total production in the previous month plus new production minus production lost to depletion. Consequently, the change in total production from one month to the next does not depend on the change in new production but rather on the difference between new production and production lost to depletion. Thus, even though the results show that new production starts to recover in the second half of 2016, total production is expected to continue to decline until the second quarter of 2017, when depletion falls to the level of new production, as shown in Figure 15. (In Figure 15, depletion is calculated using the smoothed rate shown in Figure 11 rather than the raw data). In addition, the monthly decline in total production is greatest not when new production is declining most rapidly (early 2015) but rather when the difference between depletion and new production is greatest (the second quarter of 2016).

In the base case, production of oil in the four key regions declines by 0.89 million barrels per day between April 2016 (the last data point) and its trough in May 2017, from 4.87 million barrels per day to 3.98 million barrels per day. That decline reflects the projection that new production of oil over that period of nearly 2.50 million barrels per day (a monthly average of nearly 200,000 additional barrels per day for 13 months) will fall short by 0.89 million barrels per day of the 3.38 million barrels per day of production lost through depletion over that period. The fall in base case production is projected to be most rapid during May through July of 2016, declining an average 130,000 barrels per day each month.

Total production of oil is projected to grow after mid-2017 as new production outpaces depleted production. In the base case, production regains its March 2015 peak of 5.35 million barrels per day in October 2019 and reaches 6.17 million barrels per day by the end of 2020. Production increases by 0.73 million barrels per day in 2019 (measured December to December) and by 0.68 million barrels per day in 2020. Those increases in barrels per day of production are similar to the average gain of 0.76 million in 2011-2013 but short of the record 1.27 million increase in 2014. Base case increases in production during 2019 and 2020 are similar to those in 2011-2013 despite much lower rig counts because production per rig is expected to be much higher in 2019 and 2020 than in 2011-2013.

8. Scenario Analysis

8.1. Price Scenarios

The effect of six different price paths on oil production is examined using the base case projections for technology (*epi*200, *ipr_tech*, and δ_tech). Through June 2016, the paths are identical to the price path in the base case. Beginning July 1, 2016, the real futures price of West Texas Intermediate (in 2014 dollars) abruptly changes and is held constant through the end of 2020 at one of six different levels: $30 per barrel, $40 per barrel, $50 per barrel, $60 per barrel, $70 per barrel, and $80 per barrel.[13] This section considers large and abrupt changes in the price of oil not because those are the most likely alternatives to the base case but because they illustrate the gradual short-run response of oil production to changes in prices.

As discussed above, the number of active drilling rigs is highly sensitive to the price of oil (see Figure 16). Because obtaining permits and contracting and scheduling rigs takes time, the response of drilling to the change in prices occurs gradually over seven months. By January 2017, the rig count varies from just 111 for a real futures price of $30 per barrel to a near-record 1,025 for a real futures price of $80 per barrel. (The rig count in 2017 for a futures price of $80 per barrel would be about the same on average as it was in 2014 when futures prices averaged $88 per barrel because of improved productivity.) No matter what the scenario, the rig count would rise between 2017 and 2020 as improving technology makes more fields profitable.

The response of oil production to the change in prices is both slower and proportionately smaller than the response of the rig count to the change in prices (see Figure 17). The response of production is slower for two reasons. First, because a well takes two months to be completed after it is drilled, new production only begins to respond two months after the rig count begins to respond. Second, new production is only a fraction of total production.

---

[13] To translate prices in 2014 dollars into nominal prices, multiply by 1.045 in 2017, 1.065 in 2018, 1.085 in 2019, and 1.109 in 2020. For example, in the $60 per barrel scenario, the average nominal futures price in 2020 is $60 times 1.109, or $66.54 per barrel.

Figure 16.
Rig Counts Under Alternative Scenarios for the Futures Price of Oil
Monthly Averages of the Number of Active Rigs

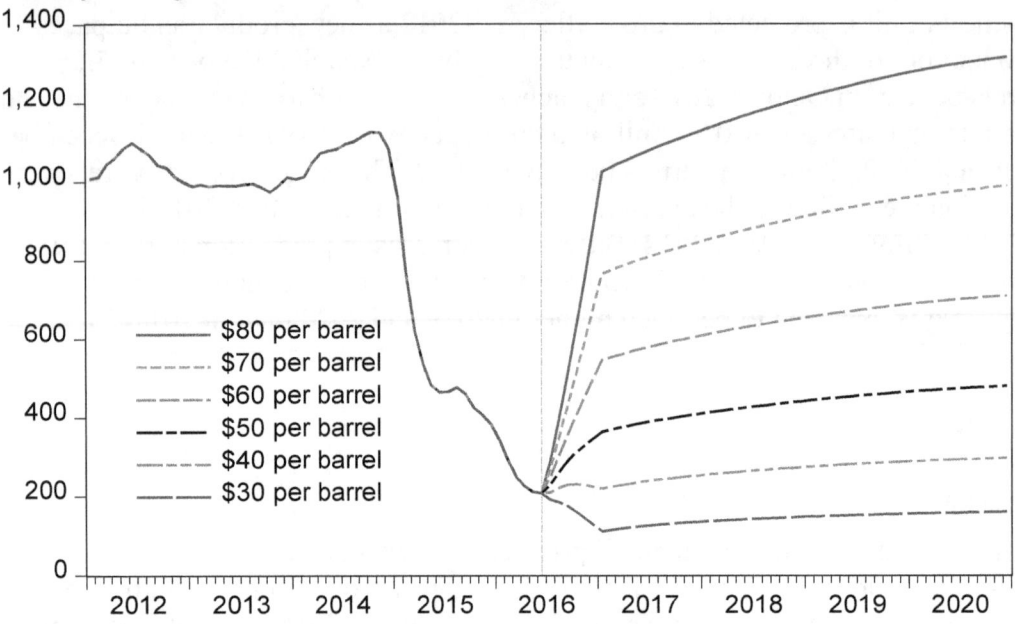

Source: Author's calculations.

Prices per barrel are in 2014 dollars and begin on July 1, 2016.

Figure 17.
Oil Production Under Alternative Scenarios for the Futures Price of Oil
Thousands of Barrels per Day

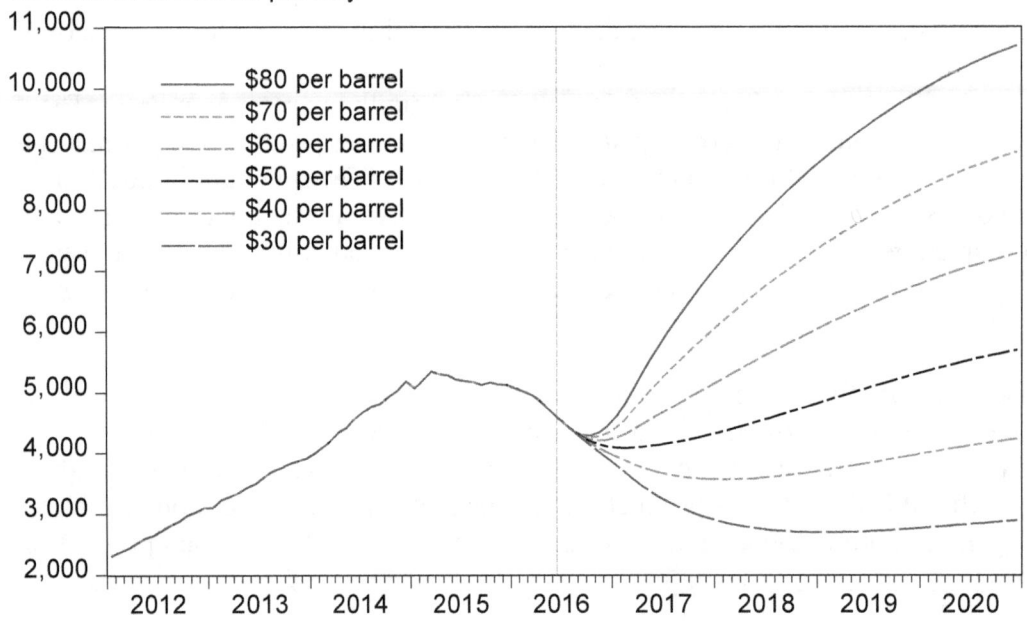

Source: Author's calculations.

Prices per barrel are in 2014 dollars and begin on July 1, 2016.

Figure 18.
Difference Between the $50-per-Barrel Scenario and the $60-per-Barrel Scenario
Percent

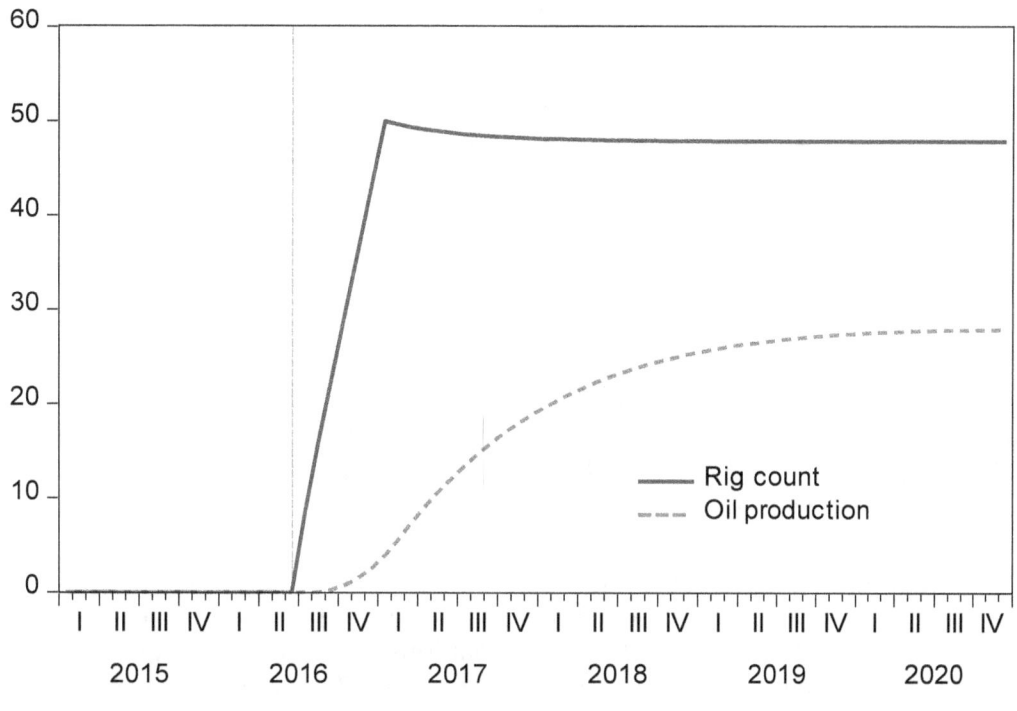

Source: Author's calculations.

The response of production is proportionately smaller than the change in the rig count because average production per rig declines as less productive opportunities are exploited. For example, the least efficient well at a price of $80 per barrel needs to be only half as productive as the least efficient well at a price of $40 per barrel to ensure that drilling at that less productive location is profitable because the oil that location produces is twice as valuable.

The time pattern of the responses of the rig count and oil production to a change in oil prices can be seen by comparing the time paths of the rig count and oil production in the $50-per-barrel and $60-per-barrel scenarios. The oil price in the $60-per-barrel scenario is permanently 20 percent higher than in the $50-per-barrel scenario beginning in July 2016.

By the end of January 2017, seven months after the change in prices, the rig count would be 50 percent higher in the $60-per-barrel scenario than in the $50-per-barrel scenario (see Figure 18). As discussed above, the response of oil production is both slower and proportionately smaller. In 2020, oil production is 28 percent higher with oil futures at $60 per barrel than at $50 per barrel.

8.2. Lessons from the Price Scenarios

One lesson from the price scenarios is that, because of the slow initial response of production to changes in the price of oil, tight oil does not reduce the short-term volatility of oil prices in response to supply shocks. The change in oil prices associated with a sudden drop in the global supply of oil would evoke no offsetting rise in shale production within the first few months after the drop. Because drilling depends on futures prices rather than on spot prices, there would be no response at all if a change in spot prices were not expected to persist.

Expanding the analysis to take uncompleted wells into account would do little to change that lesson. First, uncompleted wells take time to finish, so production from those wells would not begin until sometime after the shock. Second, because it only makes sense to leave high-cost wells uncompleted, the rise in prices needed to justify completion would lead to far more production from new drilling than from uncompleted wells.

A more important lesson from the price scenarios is that the large eventual response of tight oil production to changes in oil prices reduces the medium-term volatility of oil prices in response to supply shocks. In the scenarios, a sustained $10 per barrel increase in futures prices beginning in July 2016 would boost U.S. production of tight oil within two years by a range of 0.9 million barrels per day (going from $30 per barrel to $40 per barrel) to 1.2 million barrels per day (going from $70 per barrel to $80 per barrel).

A corollary to that lesson is that, absent a major supply shock, futures prices are unlikely to remain outside a range of $35 per barrel to $70 per barrel (in 2014 dollars) for a sustained period over the next five years. Given the historical relationship between spot prices and futures prices, that corresponds to a range of $33 per barrel to $73 per barrel for the nominal spot price of WTI. For example, in the $80-per-barrel scenario for futures prices, U.S. production of tight oil is 3.2 million barrels per day above its April 2016 level two years after the increase in prices and 4.6 million barrels per day higher after three years. It is difficult to see prices remaining at $80 per barrel with that much additional supply coming online absent a major supply shock. Even in the $70-per-barrel scenario, production of tight oil is 3.1 million barrels per day above its April 2016 level three years after the shock.

Alternately, a sustained drop in the futures price to $30 per barrel beginning in July 2016 would drive U.S. production of tight oil 1.9 million barrels per day below its April 2016 level by the end of 2017. A large increase in foreign supply would be necessary to offset that decline in production plus the increased global demand that such low prices would stimulate.

8.3. Productivity Scenarios

There are many sources of uncertainty for the future path of energy per rig, including uncertainty about technical innovations in mining and the characteristics of unexploited fields. In the model, those are represented as uncertainty about future growth of *epi*200 and *ipr_tech*. (Uncertainty about the depletion rate, which affects the timing rather than the total production of oil, can be ignored.)

The baseline projection of the futures price of oil can be used to examine two alternative sources for productivity. For the optimistic scenario, annualized growth of *ipr_tech* is

Figure 19.
Oil Production Under Alternative Scenarios for Technology
Thousands of Barrels per Day

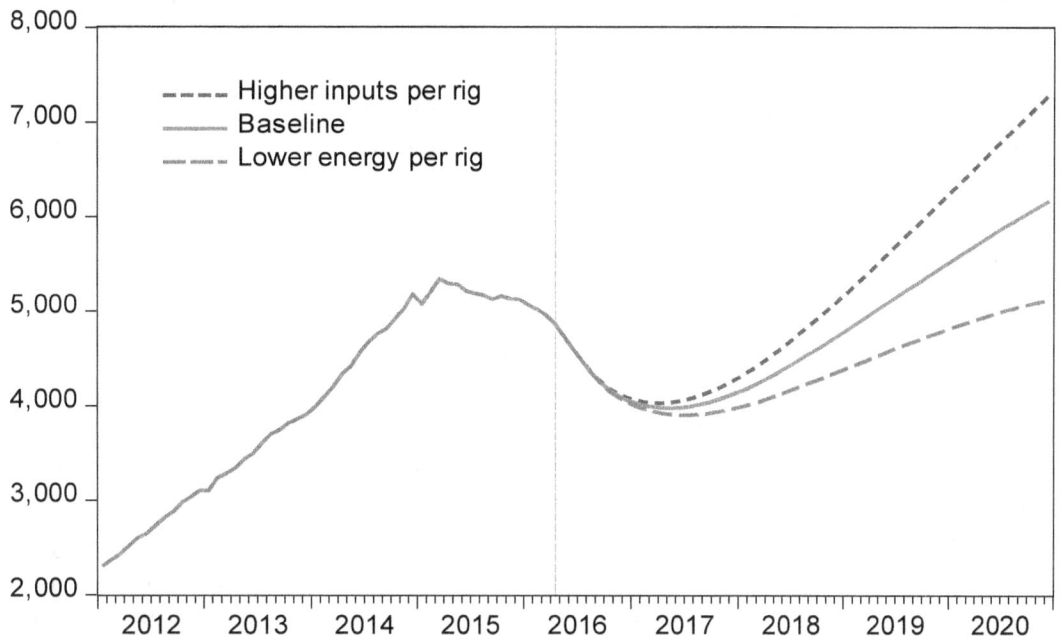

Source: Author's calculations.

Technology changes begin in May 2016.

assumed to be 5 percentage points faster than in the base case beginning in May 2016. Because the growth of *ipr_tech* is already assumed to slow to near zero by 2020 in the base case, the symmetric case in which growth of *ipr_tech* is 5 percentage points slower than in the base case is implausible. Instead, the pessimistic scenario assumes that annualized growth of *epi*200 is 2.5 percentage points slower than in the base case beginning in May 2016. That has roughly the same-sized effect on production as the larger increase in *ipr_tech* because *epi*200 affects both the rig count and energy per rig, while *ipr_tech* affects only energy per rig.

Changes in the growth of technology would have little effect in the short run but would accumulate over time (see Figure 19). The short-run impact would be small for two reasons. First, the impact on new production of oil would build slowly. An additional 5 percent increase per year is just an additional 0.4 percent increase per month, so new production would be just 0.4 percent higher in the first month of increased growth of *ipr_tech*. Second, new production is just a fraction of total production. The change in total production would grow as the cumulative change in new production grew and as new production replaced depleted production.

9. Conclusion

A fairly simple model is able to explain the rig count and production of crude oil in four key regions for production of oil from shale resources. Critical to that model are movements of drillers along an elastic supply curve that drifts down over time as mining productivity improves. While lagged effects of the drop in oil prices since late 2014 are responsible for a grim near-term outlook for U.S. production of tight oil, further improvements in mining technology lead to a good medium-term outlook. A sustained change in oil prices would have a negligible effect on production over two to three months but a large impact over two to three years. That large medium-term impact of U.S. supply would eventually blunt the impact of a supply shock occurring elsewhere.

Although data limitations prevent a quantitative analysis of mining investment in the four regions studied in this paper, some qualitative conclusions can be drawn. Mining investment in those regions should be very roughly proportional to the rig count times inputs per rig. Figure 16 shows that the lower the starting point for the price of oil, a $10-per-barrel decrease in the futures price of oil generates a smaller reduction in the rig count, and thus in investment. In contrast, each $10-per-barrel decrease in the spot price of oil boosts real consumer income by roughly the same amount regardless of the starting point of oil prices and thus probably has roughly the same effect on consumer spending. The lower the starting point for the price of oil, the greater the likelihood that the boost to consumption from a fall in oil prices exceeds the reduction in investment, causing GDP to rise.[14] There could even be a least-optimal price of oil for U.S. GDP, below which a reduction in oil prices boosts consumption by more than it reduces investment and above which an increase in oil prices boosts investment by more than it reduces consumption. In such a case, a high oil price makes the U.S. look more like Saudi Arabia, which would benefit from a further increase in oil prices, while a low oil price makes the U.S. look more like Japan or Germany, which would benefit from a further decrease in oil prices.

---

[14] David Mericle and Daan Struyven, in *Cheap Oil and the US Economy: Too Much of a Good Thing* (Goldman Sachs Economic Research, April 2, 2016), also discuss nonlinear effects of changes in the oil price on U.S. GDP.

Technical Appendix

A. Average Energy per Input and Marginal Energy per Input

Energy per input (*epi*) for all rigs active two months previously is the average of the energy per input for marginal rig j (*epim$_j$*) from a rig count (j) of zero up to the actual rig count:

$$epi = \frac{\int_{j=0}^{rigs_{-2}} epim_j}{rigs_{-2}}. \tag{A1}$$

Rearranging Equation 3 produces:

$$\frac{epim}{epi200} = \left(\frac{rigs_{-2}}{200}\right)^{-1/c1}. \tag{A2}$$

That equation implies that the productivity of the jth rig is:

$$epim_j = epi200 \left(\frac{j}{200}\right)^{-1/c1}. \tag{A3}$$

Substituting that expression for *epim$_j$* into Equation A1 yields:

$$epi = \frac{\int_{j=0}^{rigs} epi200\left(\frac{j}{200}\right)^{-1/c1}}{rigs_{-2}}. \tag{A4}$$

Solving the integral in that equation yields the expression:

$$epi = \frac{1}{1-1/c1} epi200 \left(\frac{rigs_{-2}}{200}\right)^{-1/c1}. \tag{A5}$$

Based on Equation A2, *epim* can be substituted for $epi200 \left(\frac{rigs_{-2}}{200}\right)^{-1/c1}$ yielding:

$$epi = \frac{1}{1-1/c1} epim. \tag{A6}$$

B. Methods Used for Depletion Rates

The depletion rate for production from existing wells is a function of the depletion rate for new production in each preceding month in which wells were drilled that continue to produce. The difficulty with making that calculation is that the depletion rate for new production has changed over time, so that there are as many depletion rates to keep track of as there are months in which existing wells were drilled.

In order to simplify that calculation, only integer values of monthly depletion (such as 3 percent per month or 4 percent per month) are considered. Given a depletion rate for new production, that production is assigned to the nearest two integer values. For example, if first-month production for a given month is 100,000 barrels per day and the depletion rate for new production is 3.7 percent, 30,000 barrels per day are assumed to have a depletion

31

rate of 3 percent and 70,000 barrels per day a depletion rate of 4 percent. Alternatively, if the depletion rate for new production is 4.9 percent, 10,000 barrels per day are assumed to have a depletion rate of 4 percent and 90,000 barrels per day a depletion rate of 5 percent.

Denoting *newx* as new production with integer depletion rate *x* and *prodx* as existing production with integer depletion rate *x*, production with depletion rate *x* changes over time according to the perpetual inventory method:

$$prodx = \left(1 - \frac{x}{100}\right) prodx_{-1} + newx \tag{B1}$$

where the subscript -1 denotes the prior month's value. Those values are calculated separately for oil and natural gas. Total production is found by summing over x:

$$prod = \Sigma_x \, prodx \tag{B2}$$

To go from an estimate of *ipr**δ to an estimate of *δ*, the estimate of *ipr**δ is adjusted for the estimated effect of the real futures price of oil, obtaining *ipr_tech**δ_tech. Increases in *δ_tech* are assumed to account for 80 percent of the growth of *ipr_tech**δ_tech during the 2008–2011 period and for 25 percent of the growth of *ipr_tech**δ_tech thereafter. The estimated impact of oil prices is then added back in, assuming that one-third of the impact is on *ipr* and that two-thirds is on *δ*.

To divide the depletion rate for all new production into depletion rates for new oil and new gas, the depletion rate for gas is assumed to rise one-quarter as rapidly as the depletion rate for oil. To establish levels in January 2008, *δ_tech* is set to 3.5 percent, the depletion rate for existing production of gas to 2.5 percent, and the depletion rate for existing production of oil to 2.2 percent. (A higher rate for new production than for existing production is consistent with the upward trend in the depletion rate for existing production.)